Corrugated Iron in New Zealand

Church —— Tolaga Bay

elter —— South Auckland

Credits

Photography John Maynard
Warren Viscoe
Photographs printed by Elizabeth Leyland

"The unremitting ripple" Geoff Chapple
"Dunnies should be made of tin" David Mitchell

Designer Fraser Gardyne

Introduction

This book is the work of more than ten years each, the last five of which we worked together.

Our shared interest was an affection for the material which slowly developed into a search to isolate the vernacular style of the country. We are grateful for the inspiration of our fellow travellers: anonymous builders; a small group of architects; a cow-cocky or two; and our co-authors Geoff Chapple and Dave Mitchell.

The book is a personal selection of the hundreds of photographs we have taken. It has no special order or theme, but it does show our preference for buildings rather than ephemera. We came upon corrugated iron during our holidays, on business trips and on our way to work. As our friends got to know about our project they shared their discoveries with us.

Most of the photographs are from the North Island, only because this is where we live and work. We readily admit that there is more corrugated iron of value out there in New Zealand than we know. We hope you enjoy it as much as we do.

To all those people who have given us encouragement over the last ten years we give our sincerest thanks, particularly to Elizabeth Leyland who printed our photographs.

John Maynard, Warren Viscoe

Caravan —— Ngunguru

"The unremitting ripple"

Look for the durable New Zealand image, back across our history, past what first jags the eye. Past the intended subject of the early photographs – the human bustle around the wharf sheds, the family group posed by their new house, the gold-digger reclined, with pipe, in front of his shack. And there it is, already sufficiently common that it merges into anonymity – behind them all, the unremitting ripple of corrugated iron.

Ignore it. Somehow, in these early days of the country's colonisation, it's just temporary cladding, and certainly not part of New Zealand's nineteenth century aspiration. Yet, in the twentieth century, it's still there. Not only still present, but on the land and up and down every street, the corrugated sheets have multiplied. Everywhere there's a steel sine-wave background, pervasive as black-body radiation, but still un-noticed.

Difficult to say what set the corrugated caboodle shambling down through the last century, but it probably wasn't much different from what you hear today: *Corrugated iron will do the job.* Or: *Corrugated iron makes a good roof.* The iron sheeting, a pouch of lead-head nails, and a hammer. It's our own Big Bang, an important cultural origin, and from there the ripple spread.

Nor did it stop with roofs. New Zealand is a nation of primitive builders, and like primitive builders across the world we've shaped whatever cheap material is to hand for our purposes. Corrugated iron has been slapped together for beach baches, for temporary fences which linger, for dog shelters you couldn't dignify with the name kennel . . . . The modern middle class, finding permanence and tidiness more attractive, would rather not know. But corrugated iron has now been built into our way of life, and without it the country would collapse.

Subtract the iron, wind back time in an effort to tidy the place up, and the first sheets uprooted are those already in their secondary or tertiary phase — the improvisational iron. Corrugated iron, perhaps painted hopefully in the pattern of a rock wall, which terraces the garden. Iron serving as a hotplate over the open fire, baking pipis or draining the fat in custom-made runnels from the frying sausages. Kids' canoes, the iron pinched and nailed to a wooden bow, squared at the stern, and caulked with tar off the road. Curved or bull-nosed verandah iron with 12 x 1 wooden sides, transformed into a duck-shooter's punt. The iron of tree huts, the sluice gates which change the flow of irrigation water out on the farm — wind back time and they all whirl back to the sheds, the burnt-out houses, or the derelict creameries from which they were originally poozled.

Wind back further, on this hurricane of middle-class intent, and away comes all the iron from the fences, the house roofs, the factories, the cinema halls, the hay-barns. Back into stacks.

Back to the Glenbrook mill or the smaller producers of long-run iron in New Zealand cities. Back to Japanese and Australian producers. Back to the Parnell factory of Samuel Parker in the period 1929-1886, and back to purely British sources before that.

Done. New Zealand is a nation gutted, open to the sun and the rain. Nothing but timber frames as far as the eye can see.

Not only that, but the jokes founder. Without corrugated iron as a sufficient conductor, there's no joy in hooking up the V8 coil to the makeshift urinal at rugby park. The rock on the roof makes no satisfying clang after the old man next door has shuffled to the outside dunny in his carpet slippers. The keas don't get to slide down grooves which perfectly cup their haunches, nor scrabble back to the ridging for a second run.

All this would be loss, is already a partial loss for the lack of seeing. It's time to haul corrugated iron out from under the bed. Time to admit it as the great New Zealand building material, and to recognise the way it has taken over our aesthetics.

For there never could be a cabbage tree grove without a corrugated iron shack somewhere about. Never shoulders of rolling green farmland without a half-round hay-barn. Never a bach without a corrugated iron water-tank. And more formally, across the surface of a New Zealand city, there never could be a diurnal passage of the bright New Zealand sun without the exact measure of its progress being marked by the striations of shadow on ten thousand roofs.

And if it's not that kind of day, *pik-it-a-pok*, it's the tin roof which amplifies those first drops of rain, which turns the squall into interjectory spasms of tin roof sound, or settles to a steady metallic drumming. There's not a New Zealander who hasn't huddled deeper into a warm bed at night listening to that sound. Or to the sharper *snik* of hail.

Corrugated iron is a shared link, a part of New Zealand's emotional undertow. The expatriates talk of it in London pubs, knowing more clearly from a distance where the true nostalgias lie. *Remember you'd get a stick and run flat out down the length of the fence? Yeah, or if you just walked along beside it, your footsteps had a high kind of pinging echo.*

And the heat . . . . What Zion is to the Jews, so zinc-plated iron is to the New Zealanders. It says something about their roots. There exist a million who know that intuitively, and yet there exist no formal experts. The indexes of the New Zealand history books get no closer than Coromandel, Gallipoli, but never galvanised iron. Why no honourable mention of the material that kept us dry during the impossible rainfalls of the West Coast gold-fields? That kept the ash off our heads during the great North Island burn-offs? That roofed the Maori meeting houses just as, in 1852, it roofed the first New Zealand Parliament?

The honour due and so long withheld is almost sufficient for the enthusiast to propose a flag. Simulated in silk, or the real thing, stiffly in the wind, symbol of a shared unity within our short history. And if it has to have that one identifying constellation of the southern hemisphere, why not punch it in with nails, the Southern Cross, top left? As to colour, there's no argument. Railways red.

But for corrugated iron, no flag flies. No whimper of sympathy exists from that rough public consensus which gives us our national images. For the image is a raddled one. The coat of paint is likely to be blistered. Here and there, corrosion. The iron weeps brown rust where the lead-heads have penetrated, but there's no sympathy for it. Corrugated iron might do the job, but it's not noble. Not even quaint or picturesque. It's not hewn, sawn, tamped or sun-baked like the material of vernacular architecture elsewhere in the world. It refuses to degenerate romantically into broken arches or stone ruins, or to slump quietly, like old adobe, back into the earth. It has no interstices to nurture the roots of a redeeming moss or creeper. It doesn't weather, it rusts. It doesn't tumble with the dignity of broken weight. It's light. It leans. It remains jagged and sharp, the abandoned corrugated iron shed in the field is unlovely, an old brown witch of a building, transfixed in the last spell of deserted hope and emptiness, its thin sheeting tugging at the nail.

In its final phase, it speaks only of a defeated tin-pot industrialism. The technology of stamped tin, and even during the Victorian era when it was first produced and when such simple industrial process was a source of pride, the manufacturers must have wondered who'd take it.

Rush Ho! Rush Ho! The New Zealanders did. The gold-fields of the late 1850s and 1860s produced the first big demand here. Before that, use of the new material was fitful. The Marsland Hill Barracks, built in 1855 overlooking New Plymouth to house Imperial troops, was one of the earliest. One wing of the all-iron structure still stands. In the Auckland of 1855 also, there was talk of importing an 'iron house' as the registrar's office at the Mt Eden stockade. But it was in the gold-fields and their feeder towns that the demand for speedy construction, the lack of labour for such unrewarding work as sawing planks and splitting shingles, and the need to transform sodden canvas towns into weather-proof shanty towns first arose. Corrugated iron, light, portable, weatherproof and easily erected, exactly fitted the bill.

Not just the miners' shacks wall-papered inside with old newsprint but the hotels which sprouted by the hundred were frequently clad in the iron. By the 1860s also, the material was being used for the greater glory of God, in structures like the Union Church in Naseby, Otago. And the North Island, in the process of being burnt off and stumped, was dotted with corrugated iron huts, their bush chimneys providing a fire-proof space wide and deep enough to cook damper and to encase a back-log which smouldered on through the night. As the new farms came into production, American pattern books came under scrutiny. In the 1870s, such American and British catalogues showed a wide variety of corrugated iron farming aids, from cattle-troughs to hay-barns, and the new farmers embraced the material with enthusiasm.

And then, in 1886, corrugated iron provided the thin barrier between life and death for many survivors of the Tarawera eruption. At Te Wairoa, far enough across the lake from the exploding volcano to escape total burial, but under a night bombardment of white-hot rock and volcanic mud, many of the locals sheltered under the iron roof of a hotel, or in tin dwellings. Clara Haszard, daughter of the Te Wairoa school-teacher Charles Haszard, left a graphic account of the family sheltering in their all-tin sitting room. Reverberating under the bombardment, bulging inward under the weight of debris and mud, the roof of the annexe eventually collapsed when the rafters broke, but three of the family survived.

In New Zealand a potential mythology of corrugated iron seems to await exploitation by any alert writer of the nineteenth century. Tales of survival under the volcano. Evidence of an ultimate Protestant spirit manifest in the corrugated iron churches. Tall stories from the gold-fields – the miner who tobogganed to safety on a fortuitous sheet of iron as the mud avalanche burst his tin shack, buried his mates. That other great source of myth, the railroad, begins to push through, its sheds providing a market for Samuel Parker's New Zealand-made corrugated iron, heavier duty in its zinc coat than the imported counterpart.

A mythology lies latent, yet you search in vain for a writer of the period who sees it. Thomas Bracken, one of the earliest epic versifiers, is fond of soaring the length and breadth of New Zealand in such poems as *God's Own Country*. As he dips and dives at will, somewhat in the manner of his American contemporary Walt Whitman, one awaits the glint of new roofs in the city, some homage to the inventiveness in the countryside, some worthy advance on *Leaves of Grass*, perhaps *Sheets of Iron*. But the disappointment is total. Blinded by supernal vision, by natural grandeur and the aspiring ideals of nationhood, Bracken sees no huts. Even when the subject of farming is specific, as in *The Colonist*, the "rural music" Bracken hears is never rain on the roof. Nor mention of a bush chimney. Hopelessly sentimental, it's the smoke wreathing from the chimney which catches his eye, a spiral on which "fireside prayers might mount to heaven". New Zealand is still, for Bracken, a little slice of England under the eye of an autocratic God, not a rough democracy of tin.

In any of the literature up to the turn of the century the eye of the novelist and the poet is resolutely closed to corrugated iron. The mention is sparse at best, like that of the Polish novelist, Sygurd Wisniowski, who visited New Zealand in 1864-65 and described the end of a steamship voyage in his later book *Tikera or Children of the Queen of Oceania*: "At long last we observed the grey-shingled or shiny tin roofs of New Plymouth . . . ."

It's not much, but perhaps makes the point that it takes a visitor from across the water to see what is strikingly different about the New Zealand landscape and the style of its habitation. In the early 1900s it's the writer Blanche Baughan, newly arrived from England, who sees corrugated iron most clearly. She sees the blackened stumps of the burn-off, finding in it "a curious beauty, lying in the very lap of ugliness". And within that landscape in her poem *Bush Section* she notes "the crazy tin chimney" of the settler's shack.

It's Baughan who understands how integral the tin shed is to New Zealand, and who understands also the peculiar repulsion it exerts on a European consciousness. In her collection of writings *Brown Bread From A Colonial Oven* she describes an old Swiss settler meditating on a grass slope in the South Island. He has been fifty years in New Zealand but snow-capped mountains are visible, a few cows wander, one with a bell. The alpine feeling is distinct and comparisons with his native Switzerland tug at him, but the flowers are wrong, also the trees, and then comes a last vision which totally disintegrates the nostalgia: ". . . the little hut beside the pines, as Philippe's fixed gaze at last slowly apprehended it, revealed itself most plainly to be wrought of no dear picturesque dark timbers, but of galvanised iron only, pale and ugly."

The disenchantment is swift. The old man feels fifty years of New Zealand life pop meaninglessly, feels "... that he had made a great mistake, that he had somehow been duped, defrauded, ruined".

New Zealand artists of the nineteenth century, determined not to be duped, defrauded or ruined by corrugated iron structures, solved the problem by simply not looking. In the early documentary lithographs of New Zealand cities, the intent is an overview only, the material of colonial cladding too distant within the picture to matter. The later painters move on to majestic bush and mountain landscapes, still in thrall to a romantic European vision.

It's not until the mid 1920s that such painters as Roland Hipkins shift the focus. His painting *Backyards: Dalton Street, Napier* challenges the notion that the artist seeks only sublime beauty by compulsory travel to the Otira Gorge. His surroundings may be enough, and looking out across *Backyards*, predictably, there's corrugated iron sloping down everywhere.

The director of the Canterbury College of Art, Archibald Nicoll, follows through in the early 1930s with such subjects as *Mrs Middleton's Shed*, where the overlap of the corrugated iron is visible. And Nicoll's pupil, Rata Lovell-Smith, confirms the new freedom to paint New Zealand structures with pictures like

Woolshed, Mt Burke (1932). But still, lambs gambol in the foreground, and Nicoll is more fond of autumnal light and haystacks than sheds. The paintings of both the Canterbury artists are broadly stroked, still under the sway of Impressionism. The revolution in seeing still awaits unromantic practitioners with an eye to match what A.R.D. Fairburn called, in 1934 "... the natural bleakness of our man-made scenery —buildings, bridges, railway stations and cuttings, telegraph poles...."

It was the depression, and New Zealand's attitude to itself was at a clear turning point. In 1931 the conservative parties had successfully used the slogan "Keep in step with the Motherland". But by 1935 the Labour Party won power with a programme of self-reliance and the development of indigenous resources.

The political change in the Depression years was matched exactly by a group of painters who didn't hesitate before a semi-industrial landscape, who didn't feel defrauded by tin sheds, and who sought their subjects within a framework influenced by the emerging socialism. The immigrant English artist Christopher Perkins was their articulate leader, demanding that artists "turn the stark facts of life in this country, such as the tin dairy shed, into a unit of living design". And himself painted, in *Taranaki* the bold red roof and tin water tank of a dairy factory, with Mt Egmont behind.

In literature too, corrugated iron begins to win a natural

mention. It's there in Frank Sargeson, in Ronald Hugh Morrieson. A later generation of writers picks it up. Janet Frame has a poem about a nephew who has to sleep in the basement but puts a sheet of corrugated iron outside his window to recapture the childhood sound of rain on the roof. James K. Baxter, Hone Tuwhare and Sam Hunt all strike words against the corrugated iron. Barry Crump drags his first bloody deer-skins back to the hunter's shelter, and his illustrator, Dennis Turner, opens Crump's first book *A Good Keen Man* with a drawing of that good tin shack.

In art, Eric Lee-Johnson, again with the clear vision that a return from abroad gives, transfixes the dead trees and back-blocks shacks of Northland in the 1950s. The artistic eye is wide awake by then, the path broad enough for a host of realists to depict every corrugation, for Brent Wong in the late 1960s to extend the image, his corrugations minutely detailed, stained with rust, sharp and clear under a surrealist sky. And in 1980-81 for Ralph Hotere to go right to the black heart, seizing second-hand iron itself, flicking it with paint, dribbling paint down the runnels. One of Hotere's sheets, *Aramoana: Pathway to the Sea*, sold to the Rotorua Art Gallery for $850, complete with rust holes. It's a distant cry from the last century when the iron came in from England at three shillings a sheet. Another New Zealand entirely.

And yet, not quite. Recognition is not acceptance. The clearer recognition of New Zealand as a tin-shack development area has brought its own backlash. The heir to the averted gaze of those nineteenth century recorders is the modern middle class. If anything, the prejudice against the iron has deepened. In the latter part of its history, the country as a whole has experienced rising wealth, and remains anxious to signal its advance with the appropriate symbols. Corrugated iron — still with the stigma of a poor-man's building material, still thought of as somehow temporary — is not such a symbol. From the areas of middle-class occupation, at least, it has been driven out.

But corrugated iron still matches a multitude of Kiwi needs, from a good roof down to the uses conjured by our basic love of improvising, and there's no need yet to mourn the passing of an era. The old iron continues to drift down through those levels of improvisation until it's just iron in the soil. And at the other end of the cycle, there's more iron on order, maintaining the traditional uses, and, as this book shows, for innovative use by a new generation of architects. There's no need to mourn the passing of an era, because the old iron doesn't mark an ending any more than the new iron signals a renaissance. Rather, true to its own character of numberless repetitions, the golden age of corrugated iron is with us, temporarily, always.

Geoff Chapple

Gentlemen —— Firth of Thames

"Dunnies should be made of tin"

Nothing, of course, could have stopped the houses of Abbotsford from riding with the great landslide of 1979. But if they had all been sheathed in corrugated iron they might have remained intact as they fell, later to be sky-hooked onto new sites. As it was, the brick ones shed their skins as they slipped, for the bricks played no structural role in the houses and fell away from the wooden walls to which they had been wired as the earth began to move.

Corrugated iron, nailed to those walls, would have stayed in place, and stiffened the walls greatly. The houses would have come to rest almost square and true, if sadly angled to the liquid horizontal.

Now brick is one of our most favoured "permanent" materials, and corrugated iron is one of our most despised temporary ones, but we may value the *appearance* of permanence more than the fact of it. Brick walls are inherently unsafe in a country of earthquakes and strong winds, and despite their durability and mass we no longer hold up buildings with them. We just tack them on the outside of light timber frames. It's the pieces of 4 by 2 behind that do the work, propping brick panels that are not even waterproof and supporting immensely heavy tiled roofs of coloured concrete. No reasonable person could have dreamed up the b. & t. bungalow, but reason has very little to do with popularity.

There is unlikely to be a mass market for the corrugated iron bungalows that appear in this book. The trouble is that corrugated iron looks tinny, and tinniness is the last quality the budding homeowner seeks in a house. Although brick is useless, except as a kind of long-lasting exterior wallpaper capable of stopping a cricket ball, it is expensive and has an excellent pedigree. Corrugated iron is so useful that it can only be used with dignity in the most vulgar applications.

No wonder iron has traditionally been reserved for areas of architecture that are not on show. Farm buildings can be made of tin. Dunnies should be made of tin. The invisible side walls of nineteenth century buildings were often made of tin. But the evocative front wall of any but the most meagre building could never be made of tin. The early false-fronted buildings of New Zealand towns had a concentration of decorative effort within the thickness of the street wall, in a technique that has been summed up as "Queen Anne at the front and a meatsafe at the back". Between Queen Anne and the meatsafe there was generally a lot of corrugated iron.

Iron has always suffered the stigma of the utilitarian. From the start it was too cheap to be good, too common to be valued. No building material is more waterproof, none is more simply made, nor more rapidly laid. No roof is more sweetly scaled in texture, nor more exquisitely trimmed at edges and ridges. If, as Berger has said, "there is the beauty which expresses function and the beauty which expresses hope", then the beauty of corrugated iron is mostly of the functional kind. It has none of the conspicuous excess which is so valued in more expensive and less efficient materials — and therein lies its elegance.

The corrugated iron of today is not really iron at all, but steel sheet on which a thin layer of hot zinc has been deposited. The zinc layer, which is slow to corrode, protects the steel, which would otherwise rapidly rust in the rain. In the words of the best text on tin, the Lysaght Referee, "The LIFE of a GALVANIZED STEEL SHEET is dependent upon the WEIGHT and PERFECTION of its zinc coating."

Most people who have not waved a paper fan will have thrown a paper dart, and will therefore know that a limp sheet of material can be stiffened by folding ridges and valleys into it. The stiffening ridges of corrugated iron are formed simply by passing the flat sheet between deforming rollers, which press it as a laundry wringer presses a wet towel. If, as it emerges from the rollers, it is deflected progressively by an extra roller, the sheet becomes curved permanently along its length.

Then it can be used for tank-making or laid in a long-spanning arch, in the manner of the old shop verandahs, or the roof-vents of the Chemical store on page 44 . Some of the early Nissen huts were of curved sheets of very heavy corrugated iron, which were so strong when they were bolted together at the laps that they sprang vertically from the ground and arched clean across the width of a double garage with no supporting structure at all.

There are now many rolled profiles of metal roofing available, but of them all, plain old corrugated iron has a special grace. The sine-wave curve of the corrugations has a unique structural integrity, disposing within the tiny arches of its form the loads placed on it by wind, and boots, and the heads of nails. The ridges are just high enough to keep water in the valleys, allowing adjoining sheets to nest at the edges in a weatherproof joint. In 150 years no-one has come up with a more direct profile — though with our present penchant for skimping, we now roll iron with less pronounced corrugations, drawing a little more width from each sheet and thus rendering obsolete the sheathing of thousands of existing iron buildings.

Old-profile iron was first rolled in British steel mills around 1830 and was immediately unpopular for any but the most mundane uses. It was seen as suitable for temporary and inaccessible buildings, and was exported to the colonies in great quantity. Iron was light, and could be very economically stacked in the holds of ships. It was strong and portable, beautifully suited to making instant settlements, and much in demand for building temporary cities around the goldfields of California, Australia and New Zealand. For a hundred years it remained the first choice of material in places where buildings were required quickly or natural materials were hard to come by.

In nineteenth century New Zealand, laboriously made timber shingles provided the only waterproof alternative to iron for most ordinary uses. There were no indigenous roofing materials worth using until the middle of the twentieth century, by which time a quilt of corrugated iron had covered the country, stitched from sheets that were mainly made in England or Australia, or rolled here from imported steel.

Corrugated iron may have been rolled by R. & T. Haworth of Dunedin from 1864, or by Thomas Crompton, who was established in Christchurch in 1869. But Samuel Parker, an English plumber, probably set up the first galvanising plant for roofing iron in New Zealand. He began galvanising and rolling the corrugations into iron sheets in Victoria St, Auckland, in 1886, and a few years later built a substantial factory for the purpose in Mechanics Bay. The factory, built of Parker's product, is still partly intact in 1983, sandwiched between The Strand and Augustus Terrace off Parnell Road. The Southern Cross Galvanised Iron Manufacturing Co Ltd, which Parker managed, imported English "black iron" stock and made a wide range of galvanised ware from it, including spoutings, downpipes and ridging, water tanks, tubs and buckets.

The patent "Acme" roofing nail made by Southern Cross was a forerunner of the springhead nail commonly used today in place of the expensive lead-head. The "Acme" was an ordinary wire nail fitted with a washer that stuck to the nail head as it was galvanised, acting as a rainhat to the hole the nail made in the roofing. Parker's company won a gold medal for its products at the Auckland Exhibition of 1898. And the then Minister for Public Works, R. J. Seddon, wrote a flattering testimonial to the "exceptionally good quality" of the Southern Cross brand, and authorised its use on all Government work.

New Zealand has not yet made corrugated iron from scratch, though the iron-making potential of black west coast beach sand was known when Parker opened his works. A year later the *Handbook of the New Zealand Mines, 1887* noted the wide distribution of ironsand, saying, "It remains to be proved, however, whether it can be properly treated." Attempts to extract iron from the sand began in 1848, but all were doomed to technical or economic failure until 1970, when the Glenbrook works of New Zealand Steel began producing iron from the sand and turning it to steel. New Zealand steel is used to make wire, nails and reinforcing rod. Ironically, the imported Japanese steel coil from which some corrugated roofing is rolled might conceivably be made of the ironsand New Zealand exports to Japan.

Most of the iron fixed in New Zealand came from England or Australia. A prominent supplier was John Lysaght Ltd. "Red Orb", "Blue Orb" and "Leopard" were the Lysaght brand names, the "Orbs" pictured in appropriate colour on every sheet of "maximum strength" or "curving" grade, the red "Leopard" stalking the more thinly galvanized "standard" sheets. In addition to fine iron, picturesquely branded, Lysaght produced for many years a tiny textbook on his products and their uses that was available in Australia and New Zealand.

The *Lysaght Referee* was, as the foreword claimed, "one of the handiest and most useful reference books available to the architect, builder, farmer, plumber, student and country store-keeper". It was a dense catalogue of technical information and handy hints that might have been written by Aunt Daisy had she turned her hand to building. In the same booklet were complex descriptions of carbon steels, in "drawable rimmed grade (with) hot rolled pickled finish", along with advice on how to treat strychnine swallowers and snake bite victims. "Each copy (was), of course, issued entirely without cost and with the compliments of the company." In distributing his pocket bible to all and sundry, Lysaght acknowledged that the company's products were as likely to be fixed by amateur builders as professionals.

The corrugated iron section of the *Referee* contained an addendum for New Zealand, giving such esoteric information as the number of sheets of "Blue Orb" that could be reasonably expected in a one-ton pack. Indeed, the book gave everything the novice might need to know about corrugated iron, except how to get it out of the pack and onto the building.

That is harder than it might seem, as generations of amateur builders have learned. On a windy day, a person carrying iron over roof framing can be thrown to the ground in a gust, leaving the abandoned sheet to scythe the air like a rogue frisbee. And even when the sheet is placed, the vibration of driving the first nail can set the iron sliding on the framing, carrying anyone who sits on it over the roof edge like a reluctant tobogganist. In principle, nailing iron is similar to riding a motorbike over wet railway lines. If the nail is not struck square it will slide sideways off the ridge of the sheet and penetrate the trough, producing a leak point not even chewing gum will seal.

The corners of the sheet should be avoided until the inner area is pinned down, for at the corners four sheets lap, and only a healthy whack will drive the nail through the bouncing layers of metal. Ineffectual taps with the hammer lead to exhaustion and anger. In those circumstances the nail is apt to be given a fierce clout; as it sinks home, the projecting head traps forefinger and thumb against the iron in an excruciating vice. It is time to look for the pliers.

Nailing a hand to the roof is not the only danger to be avoided. Cutting iron in long lengths with tinsnips is a lot harder than cutting cloth with scissors. The inherent stiffness of corrugated iron makes it unwilling to open at the cut line, and the hand holding the snips is easily lacerated by the cut metal edges. Further, the snips tend to wander from side to side as they cross the hills and valleys, and forcing them to hold a line increases the chance of jamming the hand in the freshly cut cleft. The skilsaw has been a boon to the amateur roofer. It screams as it cuts, but it cuts straight.

In any battle with flesh iron comes off best, but that is its only drawback. It is remarkably forgiving of bad workmanship. Even the number of laps can now be minimised, for iron has been available in full roof lengths since 1964, when N. S. Irwin Ltd installed a rolling machine that would corrugate "long-run" roofing off galvanized coil to the classic profile.

Nelson Irwin still thinks "It's the best roof. Put on properly and properly maintained it will last for 80-100 years. And it's the cheapest roof." It is said that Judge Maning's old house at Onoke, on the Hokianga harbour, has carried its original iron for 130 years, though half that time would be more common.

The value of paint in prolonging the life of a tin roof is usually unquestioned, though some argue that the galvanised zinc coating usually fails first in the invisible laps, where water is carried in by the wind and by capillary action to lie between the sheets while the exposed surfaces dry. The lapped areas are nearly always the worst protected, with only priming paint applied before the iron is fixed, while the visible surfaces are left to weather to give a good grip to the three coats of paint they usually get. And no-one, repainting a roof, lifts the laps to see how the old priming is faring. If we were all happy with the look of grey tin we might just nail it up and leave it.

Paint is the only successful decoration we have developed for iron, though for a short time around 1970 it was fashionable to spray it with hot tar and sprinkle that with "hundreds-and-thousands" of coloured sand to disguise its plebian origins.

Barn-red was the traditional paint colour. It was also the cheapest. The idea, however, that it was the easiest colour to make from the leftovers of manufacturers' paint runs is a misconception.The bubble-and-squeak of the paint-maker's kitchen is more often the colour of mud. It is true that barn-red disguises brown rust areas well, but its popularity was probably related to its price, and to the anti-corrosive red leads it once contained. Red and green roof paints were coloured more cheaply than other paints, with iron and chromium oxides that occurred naturally as earth pigments.

On the vernadah roofs of Victorian and Edwardian houses the classic red or green base colour was sometimes striped with another, it is said, to simulate the stripes of canvas awnings. If that is so, then it can be catalogued as one of the few examples in architectural history of one building material mimicking another that is cheaper and less permanent. The red and white stripes on the metal roofs of Kentucky Fried Chicken stores have a different meaning. They act as a metaphor for the striped butcher's apron we hope the Colonel once wore, forming part of an intricate visual slang that has developed rapidly to the point where red and white denote fast food, and striped roofs in any colour mean takeaway chicken. This is a new sign system that extends the long established practice of painting big advertisements on corrugated iron buildings (see pages 46, 47).

In places where aesthetic concerns are not rated, tin buildings are now rarely painted. In the country, the old red barn has given way to the unpainted semicircular shed. The green hills have been sprinkled with silver vaults. These prefabricated buildings are supremely efficient in an engineering sense, with the barest minimum of repeated elements that can be assembled with little skill. Lured by their low cost, a few people have fitted them out as houses (page 71). They lack the symbolic banality of domes and they are a lot more waterproof, but regrettably it is difficult to make private rooms inside them. They subdivide unhappily into spaces shaped like wedges of cheese, and the roof is hard to insulate or set windows into. On a hot day it acts as a radiator, and the stock sheets of translucent p.v.c. that are often let into it perform like glowing hot-plates.

It might be thought that little more could be done with corrugated iron and that it is doomed to be forever the builder's last choice, the cheapest material about, the least desired. But contemporary New Zealand architects have begun to make it a respectable material for sheathing buildings of pretension. In Wellington Ian Athfield has turned a large corrugated iron building into the Greta Point Tavern, and a few houses have been built in Auckland with iron walls under iron roofs. The house designed by John Scott (page 87) was clad in iron on a precut frame as a simple form of prefabrication.

But the current fascination of architects with iron is more aesthetic than constructional. For sixty years architects have seen themselves as functionalists, as part of the global movement of Modernism. If functionalism really had been the basis of modern architecture then we might have seen many more corrugated iron houses built in that time. Iron, after all, provides very high rewards for the effort of buying and fixing it. But the house never has been a simply practical gadget, and a complex of social pressures and aesthetic inheritance has always ridden over simply rational concerns.

Why, then, is corrugated iron making a new showing, just at a time when the principles of functionalism are under heavy attack all round the world? Surprisingly, iron has reappeared as part of an architecture based on nostalgia. The post-modern manner — avant-garde in the 1970s and fashionable in the 1980s — is full of formal and stylistic references to the architecture of earlier times. It is not the efficiency of iron that has caused its renaissance, but the visual reminders it has given us of the buildings of our past. We applaud the made-in-New Zealand look with the self-righteousness of the unconfident.

If more architects risk the aesthetic of the corrugated iron building, the old modernists among them will justify their use of tin on the ground of utility, as if aesthetic results were a mere byproduct of the exercise of common sense. Post-modernists, ransacking the old for the shock of the new, are more likely to use corrugated iron on the insides of buildings than on the outsides, and to favour the small corrugations of "baby-iron", perhaps specially rolled in stainless steel. The artist is not rewarded for slavishly accepting the old, but in the course of transforming it, he or she may leave witty historical references for the connoisseur to tick.

A post-modern transformation was performed by Auckland architectural students in 1981 when they designed and built a temporary Ironic Temple in Garden Place, Hamilton. The temple was a mock-Grecian facade made entirely of tin, save for a pair of corrugated plastic columns which lit up at night. The amused, bewildered people of Hamilton wandered about it in the evenings, discovering behind the colonnaded front and noble pediment an unholy network of scaffold props, and no building at all. What had happened to Queen Anne and the meatsafe? Was this some new kind of building that was all architecture and no use? Finally, why was corrugated iron filling the only role it had never been offered, out front and strutting with the greatest panache?

David Mitchell

Cottage —— Port Waikato

Fertiliser works —— South Auckland

Detail of opposite

Hauiti Hall —— East Coast

Plumber's store —— Nelson

Storage shed —— Nelson

Storage shed —— Hunuas

Chemical store —— Auckland

Factory —— Thames

 Garage —— Waiuku

Sheds —— Auckland

Farm shed —— Poverty Bay

Grandstand —— Sanson

Church —— Tokomaru Bay

Detail of opposite

House —— Hokitika

Garden wall —— Auckland

Entrance —— *Waiuku*

 Chimney —— Pukawa

Shot tower —— Auckland

Shop —— Kimbolton

Shed —— Wellington

House —— New Plymouth

Fuel depot —— Nelson

Bach —— Port Waikato

Detail of opposite

Facade —— Wellington

Shower — Tokaanu

Silos —— Mercer

Warehouse —— Auckland

Boat sheds —— Rawene

Boat sheds —— *Auckland*

 House —— Wanganui

Bach —— National Park

Drying shed —— Nelson

Changing sheds —— Nelson

House —— Pakiri

Railway sheds —— Hikurangi

Tree hut —— Auckland

Mail box —— Hokianga

Power station —— Thames

Office/warehouse —— Carterton

 Warehouse —— Auckland

Offices —— Wellington

 Chimney and tank —— Taupo

Cottage —— Portland

Hay shed —— Tongaparutu

Woolshed —— North Auckland

Colonial house —— Wellington

House —— Auckland

Beach shelter —— New Plymouth

Fire escape —— Auckland

Dog kennels —— Fox Glacier

Shop (rear) —— Rawene

Shop (front) —— *Rawene*

Factory —— Auckland

Shed —— Waiuku

Fence — Otago